P9-ECA-703

Measuring Matter

Developed at
The Lawrence Hall of Science,
University of California, Berkeley
Published and distributed by
Delta Education,
a member of the School Specialty Family

1325244
978-1-60902-037-8
Printing 2 – 6/2012
Quad/Graphics, Versailles, KY

Table of Contents

The Unit

I'm here at the bakery shop to see if they have a nice long loaf of French bread. Here's one. It looks pretty long. But how long is it?

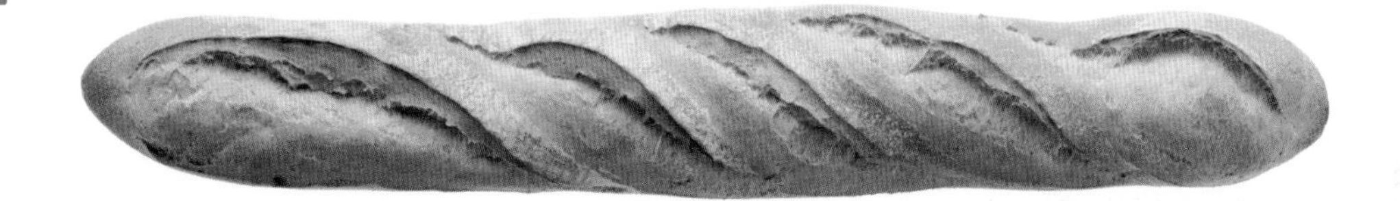

It might be as long as my pinky finger. That would be a small loaf of bread. It might be longer than a baseball bat. It's hard to tell without something for comparison.

Look, here's my pencil. I can compare the loaf of bread to my pencil.

Now I can see that the loaf of bread is longer than my pencil.

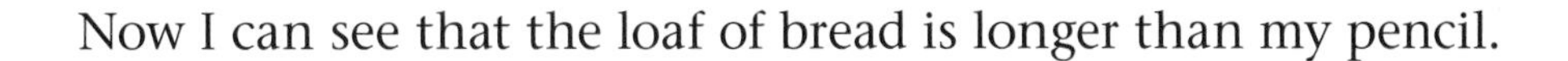

I need to tell my mom how long the loaf is. I can use my pencil as a **unit** of length. I will **measure** the loaf in pencils.

If I had a bunch of pencils, I could lay them end to end from one end of the bread to the other and count the pencils.

Now I can count the pencils. One, two, three, four, five. The bread is five pencils long. But I only have one pencil, so that won't work.

Here's an idea. I can put a mark by one end of the bread. I'll put the point of my pencil right on that mark.

I'll put a second mark on the table next to the bread to show a length of one pencil. Then I'll move the pencil forward, numbering each pencil length. The bread is five pencil lengths long.

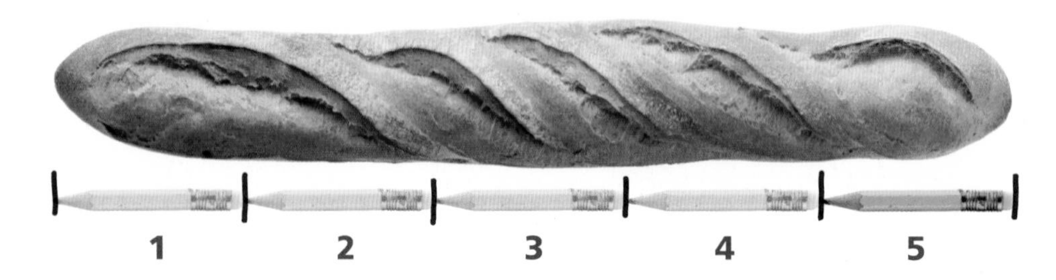

That would work, but I don't think it is a good idea to put marks on the table. A better plan would be to make a **ruler**. A ruler is a tool marked off in units. My unit is a pencil. I will make a ten-pencil ruler on a strip of paper. Numbers will remind me how many units I have measured.

Now I can easily measure the loaf of bread. The loaf is five or six pencils long. I'm not sure which line marks six, this one or this.

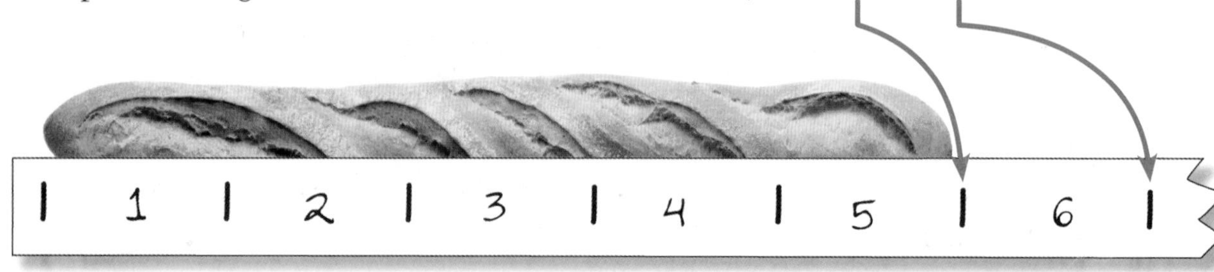

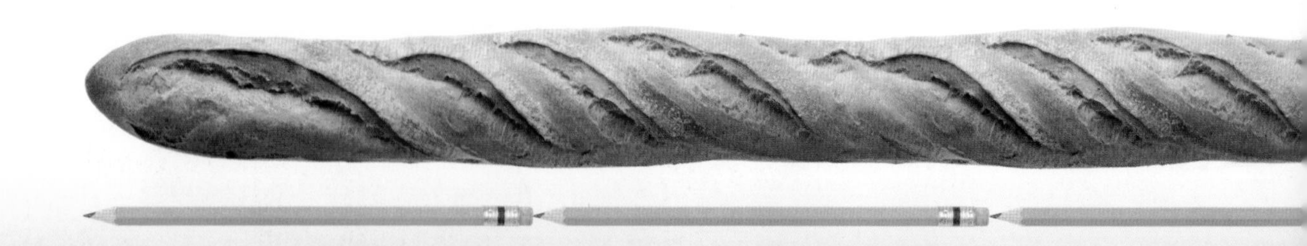

The unit "pencil" is actually the length of one pencil. I need to move the numbers from the middle of each unit to the end of each unit. I can use my first ruler to make a new one with numbers in the correct places. And I need to remember to start my ruler with zero.

Now I have a great ruler, and it is clear that the loaf of bread is five pencils long.

Later when I told my mom that I found a nice loaf of bread that was five pencils long, she was amazed. She got her pencil and measured a **distance** of five pencils. She shook her head and said, "That is way too big." When we compared our pencils, we saw the problem. Pencils are not all the same size. My mom's pencil was much longer than mine!

We need a different unit. We need a **standard** unit. A standard unit is a unit that everyone agrees to use for all **measurements**. In science, we use the **metric system** of measures. In the metric system, the standard unit for measuring distance and length is the **meter (m)**.

This is the loaf my mom imagined when I told her it was five pencils long!

The meter is a fairly large unit. The distance from the floor to a standard doorknob is about 1 meter. It is a good unit for measuring the length of a sports field or the size of a classroom.

The meter is not so good for measuring the length of pencils and loaves of bread. To measure smaller lengths, the meter has been divided into 100 equal subunits. The subunits are called **centimeters (cm)**. A centimeter is one hundredth of a meter, like one cent is one hundredth of a dollar.

Now I can measure that loaf of bread in centimeters, and my mom will know exactly how long it is.

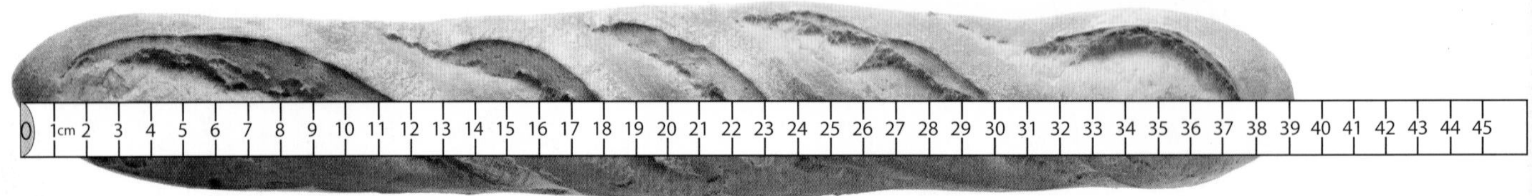

When we need to measure really long distances, like across town or the width of the state, we use **kilometers (km)**. One kilometer is 1,000 m.

Thinking about Standard Units

1. Why is it important to use a standard unit of measurement to measure length?
2. What is the standard unit for measuring length and distance in the metric system?

A Royal Measurement Mess

Many years ago, in a place far away, lived a boy named Rakim. Rakim lived with his family in a cottage on the edge of a pasture. Each morning, Rakim and his father left the cottage to tend the sheep in the pasture. Rakim's brother was a furniture maker. He made wood into beautiful chairs and tables. Rakim's sister delivered milk to the villagers. Each day, she milked and tended the family cows.

One afternoon, Rakim's sister returned from the village. She held up a sheet of paper for the entire family to see.

"Agbar, our new king, demands that every family give gifts to him," she said. "Our family must give one stone's worth of wool, two pitchers of milk, and a chair that is six hands high. We have only a week to prepare these gifts."

Over the next few days, Rakim's family prepared their gifts. Rakim and his father sheared a sheep each morning and packed the wool into bags each evening.

"Father, how will we know when we have enough wool?" Rakim asked.

"I use the stone sitting in Mother's garden to **estimate** how much to give," his father replied. "That stone is about the same size as the king's royal measurement stone. I lift up the stone, then I lift the bags of wool. When the weight feels about the same, I know we have enough wool."

After the wool was ready, Rakim helped his brother measure the chair by using a stick. "This stick is equal to the king's royal hand," Rakim's brother told him. Rakim held the stick against the chair. "It's exactly six hands!" his brother exclaimed.

The next day, Rakim and his family were to appear before the new king. Rakim's sister awoke extra early to milk her cows. She came from the barn carrying two pitchers. They were equal to the royal pitchers the king had always used.

"Those should please the king," Rakim's mother told her. "Now let's be off."

Soon they arrived at the palace. They were led to a large room where the young king sat accepting his gifts.

Rakim's father did not look happy. He was looking at the king's sour-faced gift collector. The gift collector was known as a dishonest fellow. However, the old king had made sure the gift collector treated people fairly.

Rakim's brother and sister placed their gifts in front of King Agbar. Finally, Rakim's father set down the bags of wool.

The king looked at the gifts. "These gifts are—" he began.

"Small, indeed!" cut in the gift collector. "Are you trying to cheat your new king?" He turned to a nearby soldier. "Bring out the king's measures!" he demanded.

Shortly, two soldiers staggered in carrying a large boulder between them. Another followed with a stick. One more man came in holding a colossal container.

"Here is the measure for the new king's hand," the gift collector barked, pointing at the stick. "And here are his stone and his pitcher." He pointed at the huge rock and the enormous jug. "Take the head of this family to the dungeon," he ordered.

Rakim became very angry. "Wait! This is *not fair!*" he shouted.

"Your gifts are too small!" said the gift collector. "The new king uses new measurements."

"These new measurements are much bigger than the last king's!" Rakim said. "None of the people in the village will be able to give the king his gifts using these measurements!"

"Enough!" shouted King Agbar. He turned and stared at Rakim. Finally the king spoke.

"Rakim is right. These measurements are unfair." The gift collector's jaw dropped. "From now on, Rakim will be my new gift collector." He looked at the amazed boy. "You will create standard units that will be used in this kingdom for all time. We will need a standard stone for weighing, a standard hand for measuring length, and a standard pitcher for measuring volume. Place copies of these measurements in the center of the village for all people to use. From now on, all measurements will be fair and unchanging."

So Rakim, the new royal gift collector, found a stone that was heavy, yet not too heavy for one man to lift alone. Next, he cut a stick that was exactly the length of Agbar's hand. Then Rakim used his sister's largest pitcher to measure liquids.

And that is how, many years ago in a place far away, standard measurements came to be created.

Measure This!

It is difficult to estimate the size of some things because they fool your eyes. Things that fool your eyes are called optical illusions. Look at each optical illusion below. First, answer each question just by looking at the pictures. Then use a meter tape to measure each optical illusion to check your answers.

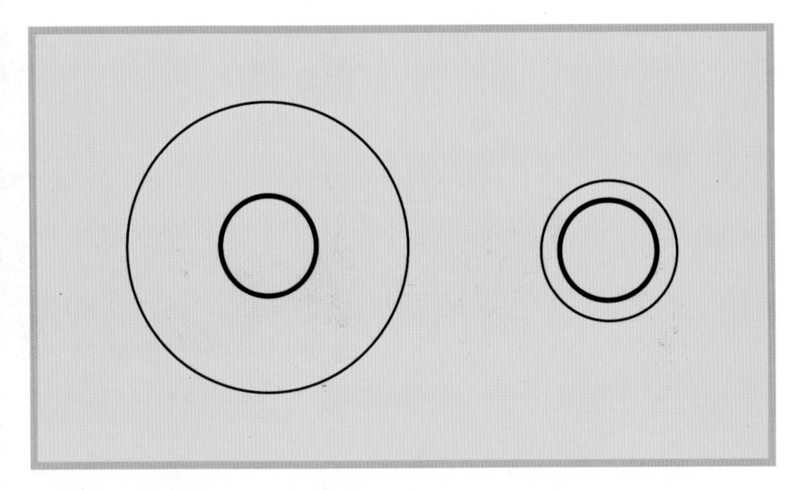

Which inner circle is bigger?

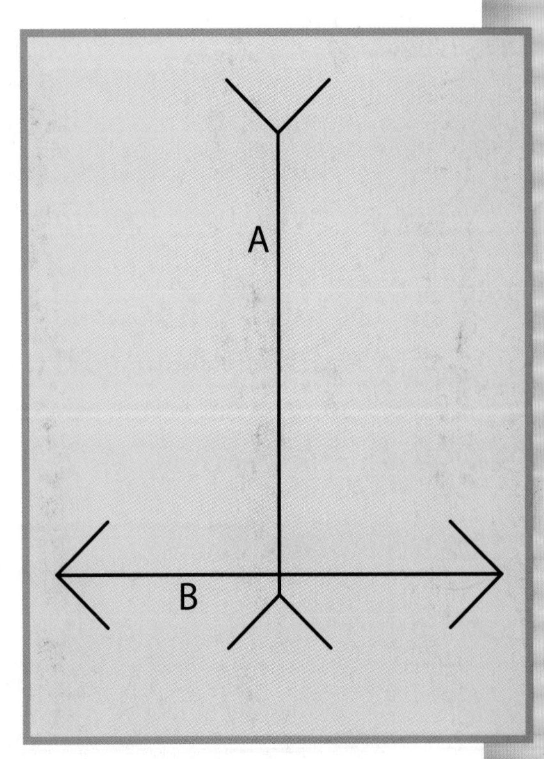

Which line is longer, A or B?

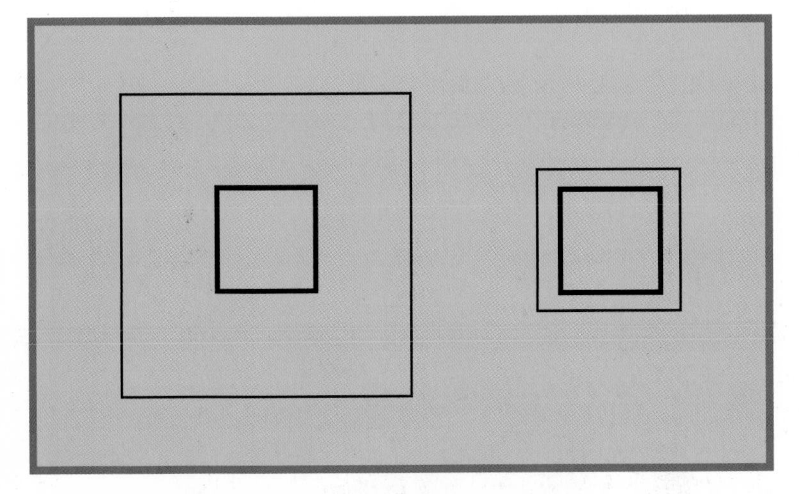

Which inner square is bigger?

States of Matter

Just about everything you see is **matter**. Whatever you are standing on is matter. Everything you are wearing is matter. Everybody you know is made of matter. All those things you eat, drink, and breathe are matter. Matter is the stuff that everything is made of.

Solid Matter

Matter is found on Earth in three common forms or **states**. Matter can be hard like bricks, flashlights, and spoons. Bricks, flashlights, and spoons keep their shape if you put them in a basin, on a table, or in a bag. And they will be the same shape tomorrow. Bricks, flashlights, and spoons are all **solids**. Solids have definite shape. And they keep that shape all the time.

Matter can be soft. Socks, stuffed toys, and sponges are all matter. Socks, stuffed toys, and sponges keep their shape if you put them in a basin, on a table, or in a bag. And they will be the same shape tomorrow. Socks, stuffed toys, and sponges have definite shape. And they keep that shape all the time.

Liquid Matter

Matter can be wet and shapeless like water, oil, and shampoo. Water, oil, and shampoo are not the same shape if you put them in a basin, on a table, or in a bag. And they will be a different shape everywhere you put them. Water, oil, and shampoo are **liquids**. Liquids flow or pour. Liquids have no shape of their own. Liquids take the shape of the containers they are in. The amount of a liquid does not change, but its shape does.

Gas Matter

Matter can be invisible and difficult to feel, like **air** and helium. Air and helium have no shape. You can't put them in a basin or on a table. They will drift away. Air and helium are **gases**. You can hold a gas in a bag. But the gas will change shape to fill the space inside the bag. The shape and **volume** of gases can change. Gases spread out everywhere. Air spreads out and is all around us.

Small Solids

Solid matter can be in tiny particles. Flour, salt, and sand are all solid matter. But sand in a basin looks different when you pour it on a table or put it in a bag. Sand can pour. Is sand a liquid? No, sand is a solid. The tiny pieces of sand are hard, and their shape and volume do not change.

Here's how to test a sample of matter to see if it is liquid or solid. Try to make a pile. If the matter will make a pile, it is solid matter. If the matter flows into a puddle, it is liquid matter.

Here's another test. Try to place a marble on the matter. If the marble stays on top, the matter is solid. If the marble sinks, the matter is liquid.

Solid particles make a pile, but liquid does not.

A marble sinks in liquid but stays on the surface of a solid.

Water Everywhere

It's easy to take water for granted. Water is everywhere. It's the most common substance on our planet. Water covers more than 70 percent of Earth. But just a tiny percentage, about 1 percent, of all Earth's water is fresh water that people can use.

Water is one of our planet's most precious **natural resources**. All living things need water to survive. In some areas of the world, water is scarce and more valuable than gold. Even in parts of the United States, droughts and water shortages can occur.

How Much Water Do We Use?

- Each American uses about 300 to 380 **liters (L)** of water each day.
- Flushing the toilet uses between 15 and 26 L each time, depending on the type of toilet.
- A bath uses about 114 L.
- A shower uses between 19 and 38 L per minute.
- Do you leave the water running when you brush your teeth? If so, you use from 3 to 7 L of water each time.
- A dripping faucet can waste more than 3,800 L of water each year.

Be a Water Watcher

What can you do to **conserve** water? An easy way to conserve water is to pay attention to how much water you use each day. Here are some other tips to help you become a water watcher.

- Keep a pitcher of water in the refrigerator. Then you won't have to run the faucet to get really cold water.
- Take short showers instead of baths.
- Use low-flow faucets and showerheads.
- Don't let the water run while you brush your teeth. Also, turn off running water while you soap up your hands.
- Don't throw facial tissues and other trash into the toilet. Use a trash can instead. This will stop clogs and cut down on the number of times you flush.
- If you have a fish tank, **recycle** the water by giving it to your plants. The fish-tank water is a good plant fertilizer.

A short shower uses less water than a bath.

Fill 'er Up

Get the lowdown on these amazing measurements!

- Elephants need a lot of water. They can drink from 75 to 100 L each day.
- The average American drinks about 19 L of orange juice each year. Nine out of every ten Florida oranges are squeezed into juice.
- The average American eats more than 24 L of ice cream a year!
- Camels are prized animals in the desert. They can go for long stretches without any water. A camel that has gone without water for a long time can drink 100 L or more at once.

The Metric System

The metric system is an easy system of measurement to use. Can you count by tens? Can you multiply and divide by tens? Then you can use the metric system.

Measurement systems based on multiples of ten were proposed many times in history. In 1793, people in France created the metric system. The French based this system on a unit they called the meter (m). *Meter* comes from the Greek word *metron,* which means measure.

How did the French set the size of the meter? They made the meter one ten-millionth of the distance from the North Pole to the equator. They wanted the meter to be based on a unit that would never change. Today the meter is based on how far light travels in a fraction of a second.

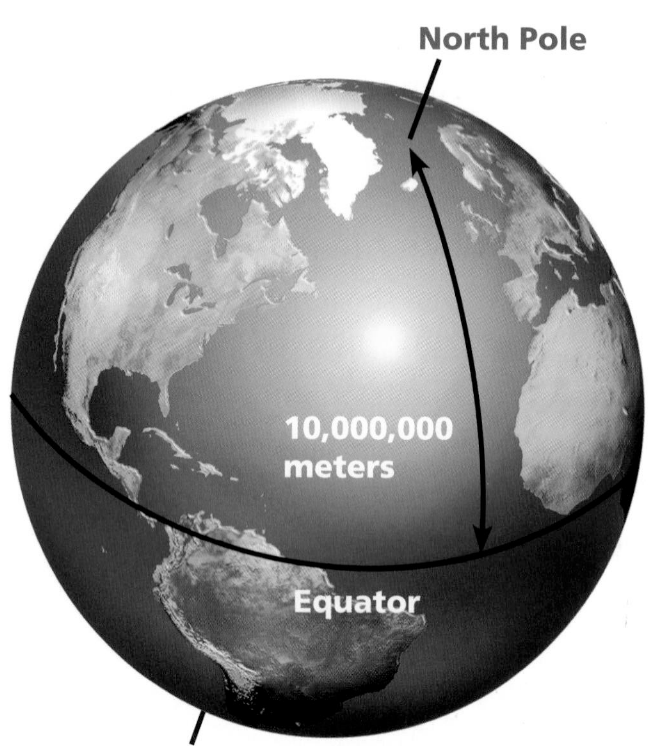

The meter was used to create other metric units. The unit of **mass** is the **gram (g)**. All matter has mass. The unit of volume is the liter (L). All matter takes up space and has volume.

Metric Prefixes

All metric units are based on the meter. The prefix can help you tell how big a metric unit is. The prefix is the part of the word that comes first.

millimeter	=	**0.001 meter** (one thousandth)
centimeter	=	**0.01 meter** (one hundredth)
decimeter	=	**0.1 meter** (one tenth)
meter	=	**1.0 meter**
dekameter	=	**10.0 meters**
hectometer	=	**100.0 meters**
kilometer	=	**1,000.0 meters**

The metric system slowly caught on around the world. Seventeen countries signed the Treaty of the Meter in 1875. This treaty created the International Bureau of Weights and Measures. The bureau adopted the metric system as the worldwide standard of measurement. Today the metric system is the standard everywhere in the world.

But the metric system is not the standard in the United States. It is the only major country in the world that does not use the metric system as its official measuring system. But even in the United States, the metric system is used in many areas. It is used in most scientific fields. It is used in many sports and recreational activities. And one day, the metric system might be used for everyday measurement in your home.

Scientists use the metric system every day.

Length, Mass, and Volume

The meter is used to define the basic units of mass and volume in the metric system. Here's how.

Mass The basic unit of mass in the metric system is the gram.

One cubic centimeter of water has a mass of 1 gram.

Volume The basic unit of volume in the metric system is the liter.

A 10-centimeter cube has a volume of 1 liter.

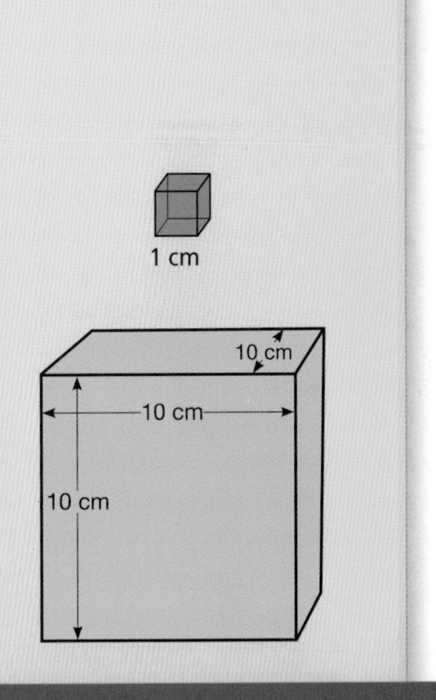

Opinion and Evidence

Two girls just finished a sponge activity. They were surprised that their 4-gram (g) sponge soaked up 32 g of water. That seems like a lot for such a small sponge.

As they recorded data in notebooks, Teasha said, "If we had a natural sponge, it would soak up even more water."

"How do you know?" asked Kim.

"I just know it would," replied Teasha. "Natural things are always better. I would always choose a natural sponge. I'm sure it would work better."

"So you've never tested a natural sponge to find out if it can soak up more water than a synthetic sponge?" asked Kim.

"Well, no, I never actually did the experiment," admitted Teasha. "But it makes sense to me that the natural sponge would soak up water better."

"We could find out for sure," said Kim. "Let's get a natural sponge and test it. That should give us **evidence** about your **opinion** that natural things are better."

A natural sponge

A synthetic sponge

The Experiment

The next day, the girls stayed after school to do their experiment. They had a new synthetic sponge and a new natural sponge. But there was a problem. The natural sponge was much larger than the synthetic one.

Teasha and Kim decided to cut three small samples from each sponge. The small samples would all be the same shape and same mass. They cut and trimmed and **weighed** carefully. Finally, all six samples were exactly 5 g.

"How should we soak the sponges to make sure it is a fair experiment?" asked Kim.

"I know," said Teasha. "We can use a stopwatch to time 1 minute while we hold the sponges under water. That will really soak the sponges. Then we'll take them out of the water. We will hold them over the basin for 30 seconds. Then we will weigh them to find out how much water soaked into each sponge."

"That sounds good to me," agreed Kim. "Let's get started."

The girls soaked and weighed the first synthetic sponge. They repeated the procedure with the other two synthetic sponges. They did this to make sure their measurements were accurate. Then they did the same thing with the three natural sponges. They recorded their measurements in a table.

The girls soaked the sponges for 1 minute.

Then they let the sponges drip for 30 seconds.

Sponge	Mass of sponge (g)	Mass of wet sponge (g)	Mass of water (g)
Synthetic 1	5	45	40
Synthetic 2	5	46	41
Synthetic 3	5	45	40
Natural 1	5	41	36
Natural 2	5	40	35
Natural 3	5	39	34

A Second Look

The girls studied the data. It looked like the synthetic sponge soaked up about 5 more grams of water than the natural sponge.

"Hmmm," said Teasha, "it looks like the natural sponge isn't better, at least not better at soaking up water. But you know what? I want to try one more thing. Let's squeeze as much water out of the sponges as we can. Then, starting with the damp sponges, we will repeat the experiment exactly. Then we will be sure our results are accurate."

Kim thought that was a good idea. They repeated the experiment and recorded these data.

Sponge	Mass of sponge (g)	Mass of wet sponge (g)	Mass of water (g)
Synthetic 1	7	45	38
Synthetic 2	8	46	38
Synthetic 3	8	45	37
Natural 1	7	41	34
Natural 2	8	40	32
Natural 3	7	39	32

"OK, I see now that the synthetic sponge is better at soaking up water," said Teasha. "The evidence is right there for all to see. From now on, I am going to use synthetic sponges to soak up spills. But I will still use natural sponges for other things because they last longer."

"Are you sure?" asked Kim.

Opinion

Teasha likes natural things. She likes chairs made of wood. She likes T-shirts made of cotton. Her opinion is that natural things are always better.

When she and Kim were working with sponges, Teasha claimed that natural sponges were better. But her claim was not based on data and evidence. Her claim was her opinion. Opinions are based on what a person believes to be true, not on scientific data. Evidence is based on observation and scientific data.

In science, claims are tested with experiments. Experiments produce data and evidence. The evidence will show if the claim is true or not true. Sometimes more experiments need to be done before a conclusion can be reached. When Teasha and Kim did their experiment, the evidence showed that the synthetic sponge soaked up more water. Teasha changed her mind about sponges after she studied the evidence.

Thinking about Opinions and Evidence

1. Teasha claimed that natural sponges were better. What did she base that claim on?
2. Why did Teasha and Kim repeat their experiment?
3. Was Teasha's claim that natural sponges last longer based on opinion or evidence?
4. What is the difference between opinion and evidence?

Vacation Aggravation

January 3

Dear Grandma and Grandpa,

We just arrived in Sydney, Australia. Wow, did I get a big surprise! I had read that the January temperatures in some parts of Australia were usually around 28°. I packed all my warmest winter clothes after I read that. When I got off the plane, it was unbelievably hot! I thought it was some kind of weird heat wave!

I made a BIG mistake. Temperatures in the northern part of Australia do average 28° in January. But that's 28°C! That's about 83°F. It turns out that because Australia is in the southern half of the world, their summer is our winter. Mom's still pretty mad. She had to buy me a bunch of shorts and shirts to wear. She says that's the last time she'll ever let me pack my own suitcase!

Love, Ami

Here's the opera house in Sydney.

January 5

Hi Grandma and Grandpa,

Today Mom and I visited the Taronga Zoo in Sydney. I couldn't wait to see the koalas.

I had read they weigh 14 kilograms. That's about 30 pounds. Did you know that koalas are endangered? Today many live in zoos.

Love, Ami

January 8

Hi Grandma and Grandpa,

Today Mom and I visited Ayers Rock. Before I got here, my friend Bill had told me that the rock was 345 feet high. I've climbed that high before, so I was excited to climb Ayers Rock. When we got here, though, Ayers Rock turned out to be 345 meters high! That's over 1,140 feet. We arrived too late in the day to climb to the top, so Mom and I enjoyed the view from the bottom.

Love, Ami

January 10

Hi Grandma and Grandpa,

Today Mom and I took a ride through the Australian countryside. We saw some kangaroos. Here's a weird fact. Australians drive on the wrong side of the road. Yep, everyone down under drives on the left. That made me nervous. When I saw that the speed limit was 100, I got VERY nervous. Then I figured out that 100 kilometers per hour is only about 62 miles per hour. After that, I could sit back and relax.

Love, Ami

January 12

G'day!

Today we checked out one of western Australia's beautiful beaches. It was terrific! And this time, I was prepared. I knew that the water temperature was a warm 20°C (about 68°F). I also knew that the walk to the beach from the hotel was 3 kilometers (1.8 miles). Now that I know that Australia, like most other countries, uses the metric system to measure things, I don't feel so out of place. Mom likes it here, too. In Australia, she weighs only 70 kilograms. (Do you know how many pounds that is?) See you in a fortnight (that's 2 weeks).

Love, Ami

Metric-to-English Conversions

These common conversions might have helped Ami while she was traveling in Australia.

- A centimeter is about half an inch.
- A meter is a little more than 3 feet, or 1 yard.
- A kilometer is about 0.6 miles.
- A kilogram is a little more than 2 pounds.
- A liter is about 1 quart.

Celsius and Fahrenheit

Celsius and Fahrenheit are two **scales** used to measure **temperature**. Both scales are based on the **freezing point** and **boiling point** of pure water at sea level. The Celsius scale has 100° between the two points. The Fahrenheit scale has 180° between the freezing point and boiling point.

Today most countries use the Celsius scale to measure temperatures. The United States, however, still uses the Fahrenheit scale.

Celsius

°C

50 45 40 35 30 25 20 15 10 5 0 -5 -10 -15

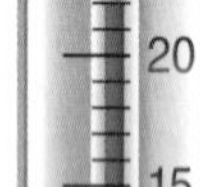

Body temp 37°

Room temp 22°

Freezing point of water 0°

Anders Celsius

The Celsius scale is named for Anders Celsius, a Swedish astronomer. Celsius lived from 1701 to 1744. In 1742, he created a temperature scale. This scale used 0°C to mark water's boiling point and 100°C to mark its freezing point. A few years later, another scientist changed Celsius's scale so that 0°C was the freezing temperature and 100°C was the boiling temperature. Celsius's scale was originally called the centigrade scale. It was renamed in the 1940s to honor the inventor.

Daniel G. Fahrenheit

The Fahrenheit scale is named for German scientist Daniel G. Fahrenheit. Fahrenheit lived from 1686 to 1736. In 1714, he invented the first mercury thermometer. He invented a temperature scale to go along with it. Fahrenheit's thermometer marked normal human body temperature as 98.6°F.

Fahrenheit thought he had found the lowest possible temperature by mixing ice and salt. He set the temperature of this mixture at 0°F. Then he set the freezing point of water at 32°F. He also set the boiling point of water at 212°F.

Fahrenheit

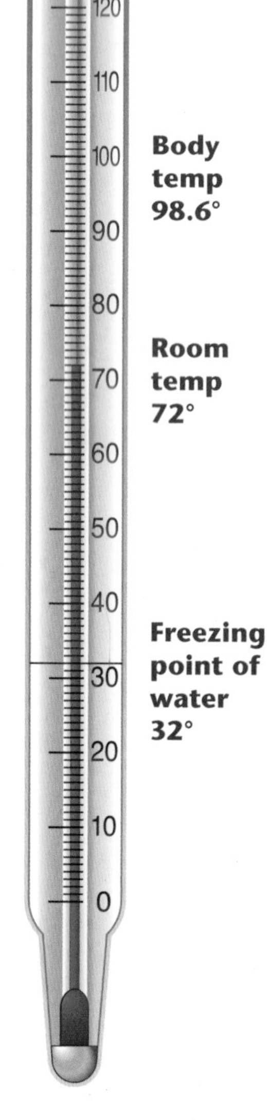

Body temp 98.6°

Room temp 72°

Freezing point of water 32°

Melt and Freeze

Ice **melts**. It **changes** from solid to liquid. An ice cube in a cup on your desk will change into water in about an hour. Chocolate and butter melt, too. But they will not melt on your desk. You have to put them in hot water to make them melt. Wax melts a little bit in hot water. It gets soft. But a pebble won't melt at all. Or will it?

What causes the butter to melt? Heat **energy**. If you put butter in a cup, nothing happens right away. If you put the cup in hot water, the butter melts. Heat energy **transfers** from the hot water to the butter. Heat energy makes the butter melt.

Heat energy is causing this butter to melt.

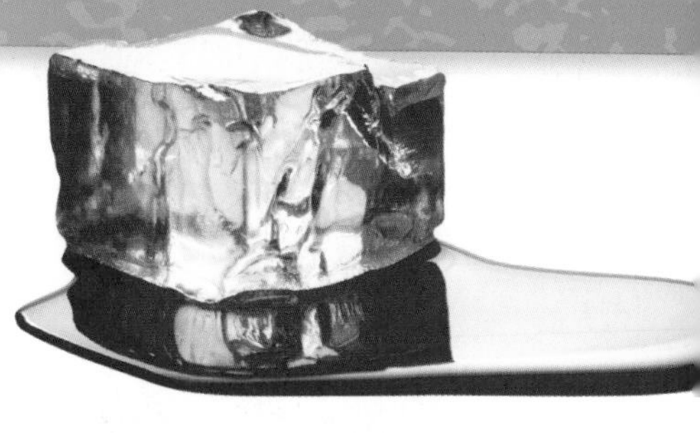

Solid ice melts to form liquid water.

But ice melts without heat energy. Why is that? Actually, heat energy does make ice melt. When ice is in the freezer, it doesn't melt. It stays solid, or frozen. When you bring ice out into a room that is warmer than the freezer, the ice melts. That's because heat energy from the room transfers to the ice and causes it to melt.

Materials melt at different temperatures. Water melts at 0°C. Water freezes at 0°C, too. When water is below 0°C, it is solid. When it is above 0°C, it is liquid. Chocolate melts at about 50°C. Candle wax melts into liquid at around 80°C. And yes, the pebble will melt when it is heated to over 1,000°C! Have you ever seen lava flowing from a volcano? That's melted rock.

Metals melt, too. Jewelers melt gold and silver to make rings and other beautiful things. Sculptors melt bronze to make statues. Iron and copper are melted to separate them from the ores taken from mines. Sand is melted to make glass. Many things that we think are always solid will melt if enough heat energy is transferred to them.

Lava flowing down the side of Kilauea Volcano in Hawaii

Gold melts at 1,064°C.

Liquid and Gas Changes

When something is wet, it is covered with water, or it has soaked up a lot of water. When it **rains**, everything outside gets wet. When you go swimming, you and your swimsuit get wet. Clothes are wet when they come out of the washer, and a dog is wet after a bath.

But things don't stay wet forever. Things get dry, often by themselves. An hour or two after the rain stops, stairs, sidewalks, and plants are dry. After a break from swimming to eat lunch, you and your swimsuit are dry. After a few hours on the clothesline, clothes are dry. A dog is dry and fluffy a short time after its bath. Where does the water go?

It **evaporates**. When water evaporates, it changes from liquid water to **water vapor**, a gas. The gas drifts away in the air.

Wet stairs just after a rainstorm **Dry stairs the next day**

What causes liquids to evaporate? Heat. When enough heat transfers to a liquid, the liquid changes into a gas. The water vapor leaves the wet object and goes into the air. As the water evaporates, the wet object gets dry.

What happens when you put a wet object in a sealed container? It stays wet. If you put a wet towel in a plastic bag, it's still wet when you take it out of the bag. Why? A little bit of the water in the towel evaporates, but it can't escape into the air. The water vapor has no place to go, so the towel stays wet.

Have you ever seen water vapor in the air? No, water vapor is invisible. When water changes into vapor, it changes into individual water particles. Water particles are too small to be seen with your eyes. The water particles move into the air among the nitrogen and oxygen particles. Water vapor becomes part of the air. When water becomes part of the air, it is no longer liquid water.

Gas to Liquid

What happens to all that water in the air? As long as the air stays warm, the water stays in the air as water vapor. But when air cools, the water particles start to come together. Tiny droplets of liquid water form.

When gas particles come together to form liquid, it is called **condensation**. Condensation is the change from gas to liquid. When water condenses, it becomes visible again. We see the condensed water as clouds, fog, and dew.

What Else Evaporates?

Water isn't the only liquid that evaporates. Gasoline for cars is a liquid. It is kept in tightly closed gas tanks. If the gasoline was left in open containers, it would evaporate and disappear into the air.

Here's an interesting fact. We breathe oxygen from the air. Oxygen in the air is a gas. But if you put a container of oxygen in a freezer that is –183°C, the oxygen becomes liquid. If there was any liquid oxygen on Earth, it would evaporate because Earth is much warmer than –183°C.

Water vapor condenses indoors, too. On a cold morning, you might see condensation on your kitchen window. Or if you go outside into the cold wearing your glasses, they will get fogged with condensation when you go back inside.

What happens to the bathroom mirror after you take a shower? The air in the bathroom is warm and filled with water vapor. When the air touches the cool mirror, the water vapor condenses on the smooth surface. That's why the mirror gets foggy and wet.

Condensation on a window

Condensation on a mirror

The Water Cycle

Water particles in the water you drink today may have once flowed down the Ohio River in the Midwest. Those same particles may have washed one of Abraham Lincoln's shirts. They might even have been in a puddle lapped up by a thirsty bison!

Water is in constant motion on Earth. You can see water in motion in rushing streams and falling raindrops and snowflakes. But water is in motion in other places, too. Water is flowing slowly through the soil. Water is drifting across the sky in clouds. Water is rising through the roots and stems of plants. Water is in motion all over the world.

Think about the Ohio River for a moment. It flows all year long, year after year. Where does the water come from to keep the river flowing?

Ohio River

The water flowing in the river is renewed all the time. Rain and snow fall in the Ohio River Valley and the hills around it. The rain soaks into the soil and runs into the river. The snow melts in the spring and supplies enough water to keep the river flowing all summer. Rain and snow keep the Ohio River flowing.

The rain and snow in the Ohio River Valley are just a tiny part of a global system of water recycling. The global water-recycling system is called the **water cycle**.

The big idea of the water cycle is this. Water evaporates from Earth's surface and goes into the air as water vapor. The water vapor condenses to form clouds. The clouds move to a new location. The water then falls to Earth's surface in the new location. The new location gets a fresh supply of water.

A simple water-cycle diagram

Water Evaporates from Earth's Surface

The Sun drives the water cycle. Energy from the Sun heats Earth's surface and changes liquid water into water vapor. The ocean is where most of the evaporation takes place. But water evaporates from lakes, rivers, soil, wet city streets, plants, animals, and wherever there is water. Water evaporates from all parts of Earth's surface, both water and land.

Water evaporates from all of Earth's surfaces.

Water vapor enters the air and makes it moist. The moist air moves up. As moist air rises, it cools. When water vapor in the air cools, it condenses. Water in the air changes from gas to liquid. Tiny droplets of liquid water form. The condensed water is visible. We see condensed water as clouds, fog, and dew.

Water vapor condenses in the air to form clouds.

Water Falls Back to Earth's Surface

Wind blows clouds around. Clouds end up over mountains, forests, cities, deserts, and the ocean. When clouds are loaded with condensed water, the water falls back to Earth's surface as rain. If the temperature is really cold, the water will **freeze** and fall to Earth's surface as snow, sleet, or hail.

Water falls back to Earth's surface as rain, snow, sleet, or hail.

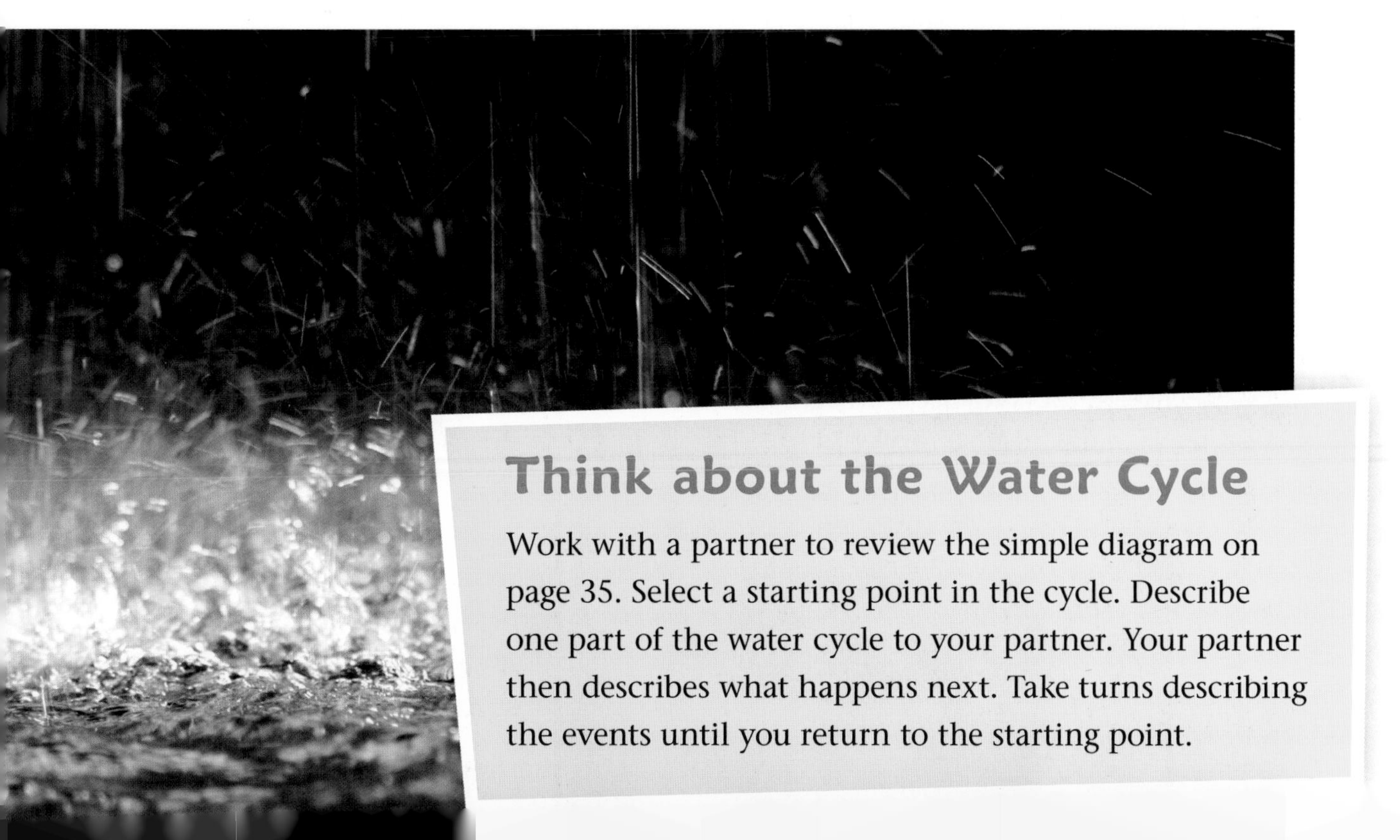

Think about the Water Cycle

Work with a partner to review the simple diagram on page 35. Select a starting point in the cycle. Describe one part of the water cycle to your partner. Your partner then describes what happens next. Take turns describing the events until you return to the starting point.

Mixtures of Solids

If you visit a lake or beach, you might see something like this at the water's edge. What's there? A **mixture**. A mixture is two or more materials together. This beach is a mixture of sand and gravel. A handful of this mixture contains bits of rock of many different sizes.

If you wanted to **separate** the gravel from the sand, how could you do that? You could pick out all the pieces of gravel one by one. But there is a faster way. You could use a **screen**. A screen has holes small enough for sand to fall through. Pieces of gravel, however, are too large to pass through. They stay on top of the screen. Screens are useful tools for separating mixtures based on the **property** of size.

A screen can separate sand and gravel.

Imagine opening a kitchen drawer to get a rubber band. Oops, the rubber bands spilled. So did a box of toothpicks and a box of paper clips. The drawer contains an accidental mixture of rubber bands, toothpicks, and paper clips. How can you separate the mixture?

You could use the property of shape. You could pick out each piece one at a time. It might take 10 minutes to separate the mixture.

Paper clips are made of steel. Steel has a useful property. Steel sticks to magnets. If you have a magnet, you can separate the steel paper clips from the mixture in a few seconds. **Magnetism** is a property that can help separate mixtures.

What about the toothpicks and rubber bands? Wood **floats** in water. Rubber **sinks** in water. The properties of floating and sinking can be used to separate the wood toothpicks and rubber bands in seconds. Drop the mixture into a cup of water. Then scoop up the toothpicks from the surface of the water. Pour the water and rubber bands through a screen. The water will pass through the screen, but the rubber bands won't. The job is done.

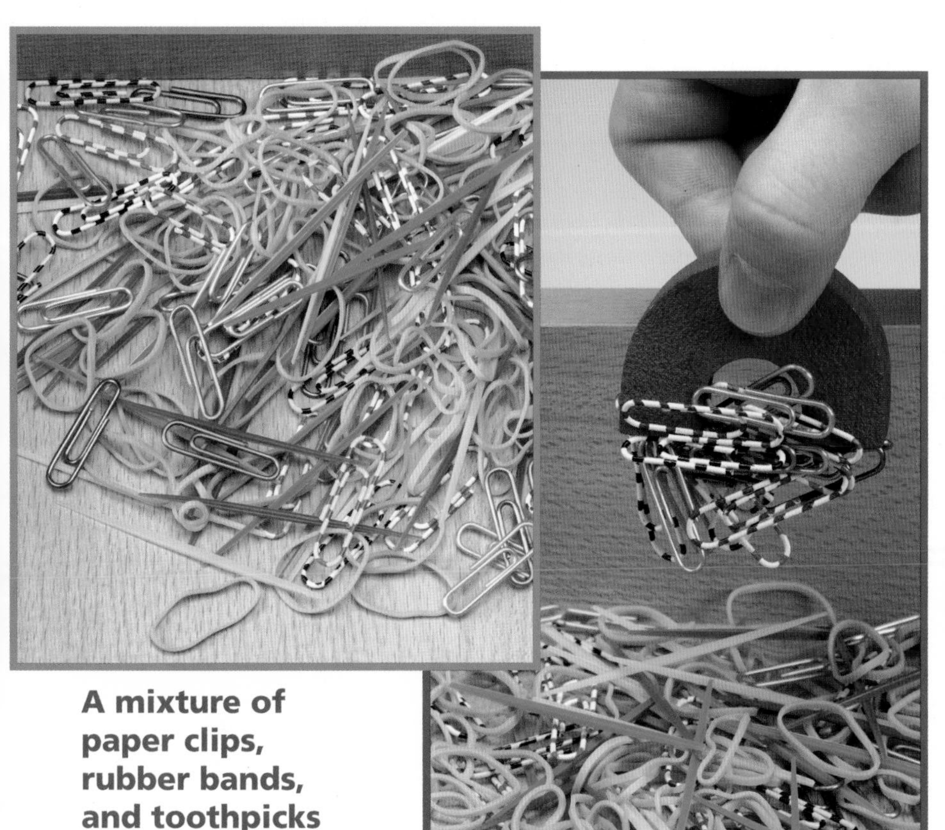

A mixture of paper clips, rubber bands, and toothpicks

Separating steel paper clips with a magnet

Separating toothpicks and rubber bands in water

Solids and Liquids

Mixtures of solids and liquids are interesting. Several things can happen. When sand and water are mixed, the sand sinks to the bottom of the container. If you stir the mixture, things move around, but that's about it.

When you mix **chalk** and water, the chalk makes the mixture **cloudy** white. After a while, the chalk settles to the bottom.

When you mix **salt** and water, the salt disappears, and the mixture is **transparent** and colorless.

Sand, chalk, and salt all make mixtures with water. After stirring, you can still see the sand and chalk, but the salt has disappeared. Salt is different in some way.

A mixture of salt and water forms a **solution**. A solution is a special kind of mixture. When solid salt and liquid water are mixed, the solid disappears into the liquid. The solution is transparent.

When the solid salt disappears in the water, it is *not* gone. It has **dissolved**. When a solid dissolves, it breaks into pieces so tiny that they are invisible. When salt dissolves in water, it makes a saltwater solution.

Mixing sand and water

Mixing chalk and water

Mixing salt and water

Sand mixture after 5 minutes

Chalk mixture after 5 minutes

Salt mixture after 5 minutes

Reactions

Vinegar and baking soda are two materials you have worked with in class. Vinegar and baking soda have properties that help you identify them. Vinegar is a liquid with a strong smell. Baking soda is a solid in the form of a powder.

Carlo did an experiment to see what happens when vinegar and baking soda are mixed. He put solid baking soda in one cup. He put liquid vinegar in another cup.

Baking soda and vinegar

Carlo put the vinegar cup inside the baking soda cup. He put the two cups on one side of a balance and mass pieces on the other side. He added mass pieces until the system balanced.

Mass pieces equal to the mass of the baking soda and vinegar

Carlo carefully poured the vinegar into the cup with baking soda. The mixture fizzed and bubbled.

What happened? A **chemical reaction**. The vinegar and baking soda reacted. During the reaction, new materials formed. One of the new materials was a gas. The gas that made all the bubbles was a new material. Where did the gas come from?

Carbon dioxide gas forms when vinegar and baking soda are mixed.

The particles in the vinegar and baking soda combined in new ways during the reaction. One new combination formed the gas **carbon dioxide**. That's where the gas came from. The gas was a new material that formed when vinegar and baking soda reacted.

After the fizzing stopped, Carlo looked in the cup. There was no solid baking soda left. He carefully waved his hand over the cup to bring the smell toward his nose. It no longer smelled like vinegar. The new materials had different properties than the starting materials.

Carlo made one more observation. The mass pieces were still in the cup on the balance. He put his two cups back on the balance. The system did not balance. The reaction cup had less mass than it did before. Why?

Gas is matter. All matter has mass. When the carbon dioxide gas formed, it went into the air. Millions of particles left the cup and went into the air. The material in the cup lost mass.

Particles combine to form new materials. Every different combination of particles makes a different material. The particles rearrange during reactions. New arrangements of particles make new materials.

Carlo made a new material, carbon dioxide gas, by combining baking soda and vinegar.

The mass in the cups after the reaction is less than it was before the reaction.

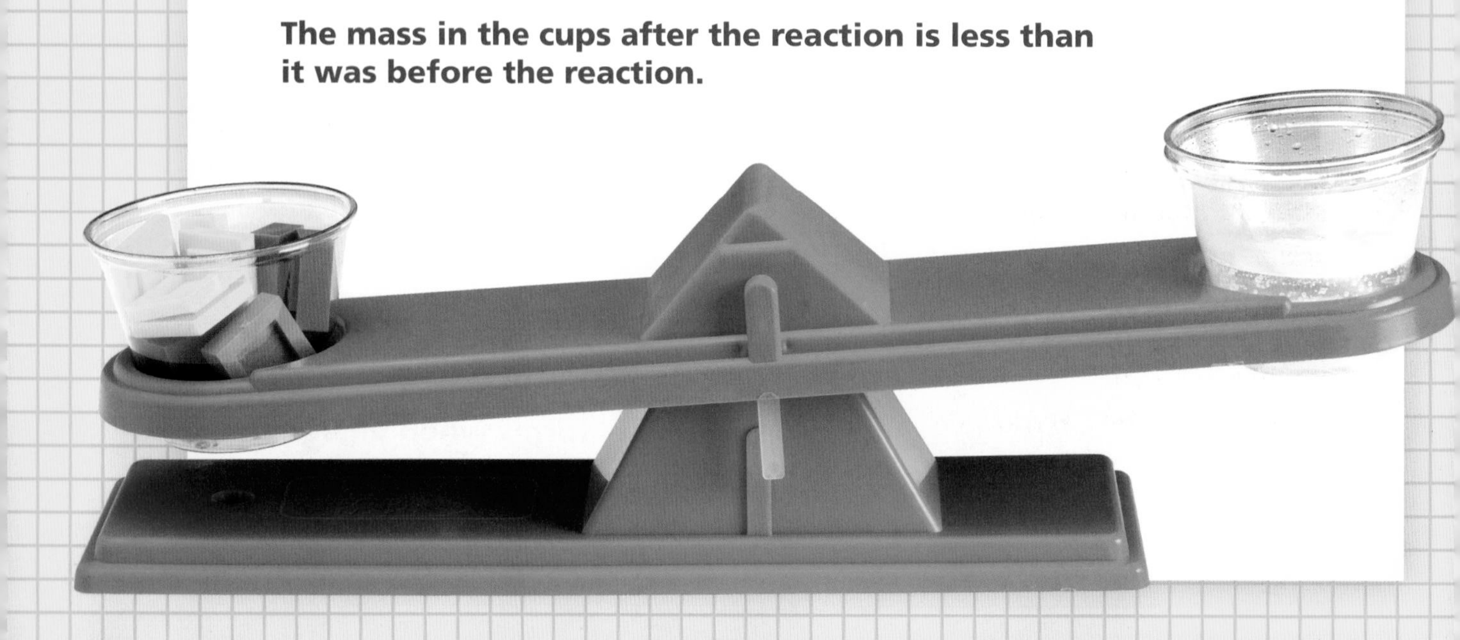

Careers You Can Count On

The metric system is the international system of measurement. The United States does not officially use the metric system. Yet every day, we count on accurate measures. We use measurements at home, work, and play. Here are a few people who use measurements at work.

Scientists

Today scientists around the world use the metric system. In the United States, scientists use it for their research. They use the metric system to measure, collect, compare, describe, and analyze information.

Scientists collect and record information using the metric system.

Pharmacists

Today most medicines are measured using the metric system. Pharmacists measure and label all types of medicines using the metric system. For example, on a bottle of aspirin, the dosage is listed in milligrams (mg).

Pharmacists carefully measure medicines to be sure the dosage is correct.

Meteorologists

Meteorologists study Earth's air and weather. They use the metric system to measure temperature and other weather conditions. They also use it to measure the amounts of chemicals in Earth's air.

Meteorologists use the metric system to predict weather conditions.

Biologists

Biologists study living things. They use the metric system to measure and weigh animals and plants. They also use metric measures to map the places where plants and animals live. Biologists who work in zoos use metric measures to help them take care of the animals. Like pharmacists, biologists measure medicines using the metric system.

Biologists measure the growth of plants and animals.

Astronauts

Astronauts are trained to take part in spaceflights. The National Aeronautics and Space Administration (NASA) uses metric weights and measures on all flights.

Astronauts use metric measurements for all their space duties.

Ecologists

Ecologists study the relationships between living things and their environments. Ecologists measure how much pollution is being released into Earth's air and water. They use the metric system for these measurements. Ecologists also use the metric system to measure and map the loss of some environments. This loss is caused by droughts, floods, fires, and natural disasters, as well as by human impacts such as climate change.

Ecologists measure water pollution in metric units.

Archaeologists

Archaeologists study how people lived long ago. They use the metric system when charting and mapping areas they are studying. They also measure and weigh bones and other objects using the metric system.

Archaeologists carefully measure and record each new find.

Chefs and Bakers

Chefs and bakers use many different types of measurements at work. They need to carefully measure all the ingredients in recipes. Bakers also need to know what temperatures their ovens must be. Then they can cook breads, cakes, and other delicious foods.

A skilled chef must measure the right ingredients to prepare a meal.

Carpenters and Architects

Carpenters and architects have to understand many types of metric measurements. They use tape measures, rulers, and other measuring instruments. They also must know how to read blueprints correctly. Blueprints are plans that tell exactly what a building will look like.

Architects use exact measurements when making blueprints.

Auto Mechanics

Auto mechanics who work on cars from other countries must know the metric system. Instruction books and packaging materials from other countries may use only metric measurements. This also is true of parts used to repair foreign cars and trucks.

Mechanics must use metric tools to repair foreign cars.

Athletes and Sports Officials

Many sports use the metric system for measurement. Track-and-field events, swimming, and skiing are just a few. Runners compete in the 100-, 200-, 400-, and 1,000-meter dashes. Cyclists compete in 10-kilometer races. Divers compete using platforms 10 meters high. That's almost as high as a three-story building!

Athletes often compete against one another in different countries. Because most countries use the metric system, it is used at international sporting events as well.

The Boston Marathon is the world's oldest and most well-known marathon. In 1975, the Boston Marathon was the first to include a wheelchair division competition. In 2004, Ernst Van Dyk (1973–) of South Africa set the world record for finishing the Boston Marathon with a time of 1 hour, 18 minutes, and 27 seconds. Van Dyk has won the marathon nine times.

Amazing Athletic Achievements

These are some world records for international sporting events.

Event	Time/distance	Record holder	Date
MEN'S TRACK & FIELD			
100-meter dash	9.58 seconds	Usain Bolt, Jamaica	August 16, 2009
200-meter dash	19.19 seconds	Usain Bolt, Jamaica	August 20, 2009
Long jump	8.95 meters	Mike Powell, USA	August 30, 1991
WOMEN'S TRACK & FIELD			
100-meter dash	10.49 seconds	Florence Griffith Joyner, USA	July 16, 1988
200-meter dash	21.34 seconds	Florence Griffith Joyner, USA	September 29, 1988
Long jump	7.52 meters	Galina Chistyakova, USSR	June 11, 1988

Veterinarians

Veterinarians use the metric system to measure medicine for animal patients. They also use temperature measurements to see how healthy their patients are.

Veterinarians use the metric system to measure the growth of animals.

Teachers

Science and math teachers help students learn about the metric system. Nearly all the nations of the world use the metric system. That is why it is important for people in the United States to understand and feel comfortable with the metric system.

Teachers explain the metric system to young scientists.

Using the Metric System

Would you believe that we use the metric system already? We use it every single day.

When we talk about electricity, we talk about watts. Watts are metric units.

We buy liters of soft drinks.

We measure medicine and vitamin dosages in milligrams.

Hunt for Metrics!

Can you find five things around your house that have metric measurements on them? Make a list on a sheet of paper.

Science Safety Rules

1. Listen carefully to your teacher's instructions. Follow all directions. Ask questions if you don't know what to do.
2. Tell your teacher if you have any allergies.
3. Never put any materials in your mouth. Do not taste anything unless your teacher tells you to do so.
4. Never smell any unknown material. If your teacher tells you to smell something, wave your hand over the material to bring the smell toward your nose.
5. Do not touch your face, mouth, ears, eyes, or nose while working with chemicals, plants, or animals.
6. Always protect your eyes. Wear safety goggles when necessary. Tell your teacher if you wear contact lenses.
7. Always wash your hands with soap and warm water after handling chemicals, plants, or animals.
8. Never mix any chemicals unless your teacher tells you to do so.
9. Report all spills, accidents, and injuries to your teacher.
10. Treat animals with respect, caution, and consideration.
11. Clean up your work space after each investigation.
12. Act responsibly during all science activities.

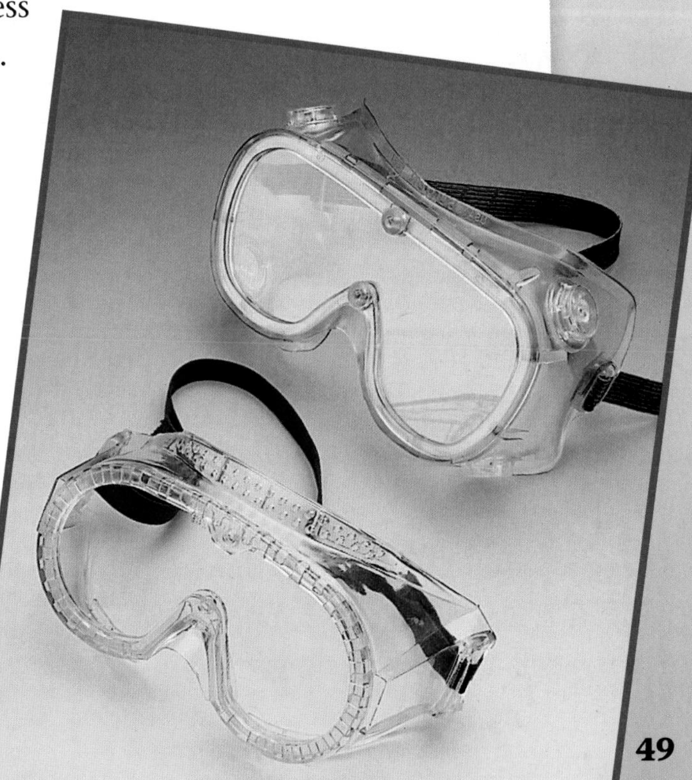

Glossary

air the mixture of gases surrounding Earth

boiling point (100°C) the temperature at which water changes to gas

carbon dioxide (CO_2) a gas made of carbon and oxygen

Celsius (°C) the basic unit of temperature in the metric system. Water freezes at 0°C and boils at 100°C.

centimeter (cm) a unit in the metric system; one hundredth of a meter

chalk one form of the material calcium carbonate

change to make different

chemical reaction an interaction between materials that produces one or more new materials that have different properties than the starting materials

cloudy not clear

condensation the process by which water vapor changes into liquid water, usually on a surface

conserve to use carefully and protect

dissolve to mix a material uniformly into another

distance the amount of space between two points

energy the ability to make things happen. Energy can take a number of forms, such as heat and light.

estimate to make an educated guess about a measurement

evaporate when water in a material dries up and goes into the air

evidence data used to support claims. Evidence is based on observation and scientific data.

float to stay on the surface of water because of being less dense than water

freeze to change from liquid to solid

freezing point (0°C) the temperature at which water becomes a solid (ice)

gas a state of matter with no definite shape or volume; usually invisible

gram (g) the basic unit of mass in the metric system

kilometer (km) a unit in the metric system; 1,000 meters

liquid a state of matter with no definite shape but a definite volume

liter (L) the basic unit of liquid volume in the metric system

magnetism a force that attracts iron and steel

mass the amount of material in something

matter anything that has mass and takes up space

measure to compare the size, capacity, or mass of an object to a known object or system

measurement the number description of a property of an object or system

melt to change from a solid to a liquid state as a result of warming

meter (m) the basic unit of distance or length in the metric system

metric system a measuring system based on multiples of ten

mixture two or more materials together

natural resource a material such as soil or water that comes from the natural environment

opinion a claim based on belief, not on scientific data or observations

property something that you can observe about an object or a material. Size, color, shape, texture, and smell are properties.

rain liquid water that is condensed from water vapor in the atmosphere and falls to Earth in drops

recycle to use again

ruler a tool, marked in standard units, used for measuring distance and length

salt a solid white material that dissolves in water; also known as sodium chloride

scale something divided into regular spaces to use as a tool for measuring. Rulers and thermometers are both scales.

screen wire mesh used to separate large and small objects

separate to take apart

sink to go under water because of being more dense than water

solid a state of matter that has a definite shape and volume

solution a mixture formed when one or more substances dissolve in another

standard something that is recognized and used in a measurement system

state a form of matter. The three common states of matter are solid, liquid, and gas.

temperature a measure of how hot or cold matter is

transfer to pass from one object to another. Heat energy transfers to make objects melt.

transparent clear

unit a length, volume, or mass used as the basis for a measurement system

volume three-dimensional space

water cycle the repeating sequence of condensation and evaporation of water on Earth, causing clouds and rain and other forms of precipitation

water vapor the gaseous state of water

weigh to find the mass of. An object is weighed to find its mass.

Index